THE
INFLUENCE
OF
MAN

MATTHEW A. GALLAGHER

"How dangerous is the acquirement of knowledge and how much happier that man is who believes his native town to be the world, than he who aspires to be greater than his nature will allow."
- Mary Shelley's *Frankenstein* pub. 1818

Introduction

Welcome to the inherent ramblings of an observer in the world around us.

This work is for those who thoroughly enjoy trying to identify the interrelation of systems – just to gain a greater understanding of the world around them.

When one walks through life with the pure intention of wanting to learn—they begin to see. So this book is meant for you to learn to see.

It should be an easy read…

It is not overly complex, nor filled with platitudes and meaningless drivel. It is clear, concise, to the point, and void of hyper-inflated [redacted].

As such, I welcome you to read this perspective so that you may too learn to see – The Influence of Man.

Table of Contents

Introduction

Chapter I. Current State of Affairs

Chapter II. Tale of Two Dogs

Chapter III. Tale of The Old man

Chapter IV. The Idiocy that is Racism, The Weakness that is Sexism, The Insecurity that is Classism, & Concepts of Tribalism

Chapter V. Reeducating Education

Chapter VI. Concepts of Climate Change

Chapter VII. Foundational Climatologic Destruction (F.C.D.) & Rates of Concern

Chapter VIII. Ideas of Ages Past

Chapter IX. Shoulders of Giants

Chapter X. The Purpose of Life

Dedications

Appendix

Before going any further with the reading of the main chapters of this book – it is important to take a step back to view our national and global landscape and better understand the world as it is today.

The following information has little to do with subsequent chapter topics – at least at first glance. But as all major problems have individual impacts, it is imperative to see the challenges around us in order that we may make a difference on an individual basis.

Let us examine... The Current State of Affairs.

Chapter I. – The Current State of Affairs

In 2016, the U.S. and the world looked like this…

The U.S. GDP was approximately $18.56 Trillion.

The U.S. National Debt is expected to surpass $23.2 Trillion by the end of 2017.

The U.S. Federal Spending for 2017 is estimated to be …

Item	Cost (in Billions)	Percentage
Social Security, Unemployment, labor	$1,327	33.77%
Medicare & Health	$1,031	27.02%
Military	$542.93	14.23%
Interest on Debt	$266.36	6.98%
Veteran's Benefits	$159.07	4.17%
Food & Agriculture	$135.70	3.56%
Education Training, Employment,	$94.65	2.48%
Transportation	$88.20	2.31%
Administration of Justice	$56.24	1.47%
International Affairs	$49.14	1.29%
Energy & Environment	$38.31	1.00%
Science	$27.72	0.73%

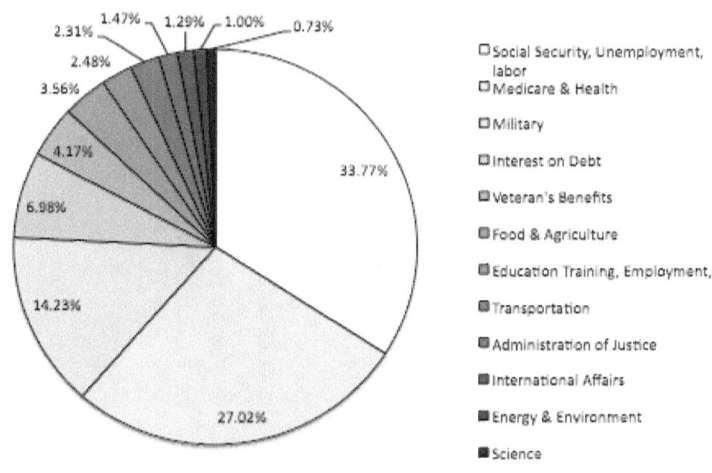

9.3% of the U.S. population or ~29.1M people have diabetes.

11.5% of U.S. adults or ~27.6M people have heart disease.

35.7% of U.S. adults or ~85.7M people are obese.

6.7% of U.S. adults or ~16.1M people suffer from depression.

18.1% of U.S. adults or ~43.4M people suffer from anxiety.

14.5% of the U.S. population or ~45M people live below the poverty line.

0.91% of U.S. citizens or ~2.2M people are in incarceration.

There are over 500,000 U.S. citizens who are homeless.

More than half of the U.S. population has less than $1,000 savings.

There are 16,000+ homicides in the U.S. every year.

There are over 15,000 nuclear weapons in the world.

795M people across the globe are undernourished.

783M people across the globe do not have access to clean water.

1.3B people across the globe do not have access to electricity.

The following species are considered **endangered**…

- *Lycaon pictus* (African Wild Dog)
- *Panthera tigris altaica* (Amur Tiger)
- *Elephas maximus indicus* (Asian Elephant)
- *Panthera tigris tigris* (Bengal Tiger)
- *Mustela Nigripes* (Black-footed Ferret)
- *Balaenoptera musculus* (Blue Whale)
- *Thunnus spp* (Blue fin Tuna)
- *Pan paniscus* (Bonobo)
- *Elephas maximus borneensis* (Borneo Pygmy Elephant)
- *Pan Troglodytes* (Chimpanzee)
- *Balaenoptera physalus* (Fin Whale)
- *Speniscus mendiculus* (Galapagos Penguin)
- *Platanista gangetica gangetica* (Ganges River Dolphin)
- *Chelonia mydas* (Green Turtle)
- *Cephalorhynchus hectori* (Hector's Dolphin)
- *Cheilinus undulates* (Humphead Wrasse)
- *Elephas maximus indicus* (Indian Elephant)
- *Panthera tigris corbetti* (Indochinese Tiger)
- *Plantanista minor* (Indus River Dolphin)
- *Panthera tigris jacksoni* (Malayan Tiger)
- *Eubalaena glacialis* (North Atlantic Right Whale)
- *Ailurus fulgens* (Red Panda)
- *Zalophus wollebaeki* (Sea Lions)
- *Balaenoptera borealis* (Sei Whale)
- *Panthera unica* (Snow Leopard)
- *Elephas maximus maximus* (Sri Lankan Elephant)

The following species are considered **critically endangered**...

- *Panthera pardus orientalis* (Amur Leopard)
- *Diceros bicornis* (Black Rhino)
- *Pongo pygmaeus* (Bornean Orangutan)
- *Gorilla gorilla diehli* (Cross River Gorilla)
- *Gorilla beringei graueri* (Eastern Lowland Gorilla)
- *Eretmochelys imbricata* (Hawksbill Turtle)
- *Rhinoceros sondaicus* (Javan Rhino)
- *Dermochelys coriacea* (Leatherback Turtle)
- *Gorilla beringei beringei* (Mountain Gorilla)
- *Ferae Pholidota* (Pangolin)
- *Pseudoryx nghetinhensis* (Saola)
- *Panthera tigris amoyensis* (South China Tiger)
- *Elephas maximus sumatranus* (Sumatran Elephant)
- *Pongo Abelii* (Sumatran Orangutan)
- *Dicerorhinus sumatrensis* (Sumatran Rhino)
- *Panthera tirgris sumatrae* (Sumatran Tiger)
- *Phocoena sinus* (Vaquita)
- *Gorilla gorilla gorilla* (Western Lowland Gorilla)

"No snowflake in an avalanche ever feels responsible."

    - Stanislaw Jerzy Lec

We begin our journey with how we perceive our reality and introspective approaches to the way we view the world around us. Our perception defines how we see, how we walk, how we act, and how we lead.

We must challenge our perceptions and conventional notions to ensure accuracy – or else we are doomed to a fate of misunderstanding and conflict.

There is no better way to show how true this is than… the Tale of Two Dogs.

Chapter II. – Tale of Two Dogs

This is the tale of a pet owner and an experience with his beloved pets. The owner has two dogs – one, a happy-go-lucky Retriever... the other, a reserved Labrador.

The owner takes both dogs to two separate doors—neither dog knows what is in the room behind the doors.

The owner takes the Retriever and walks him to the first door. The Retriever, always anticipating a walk in the park, excitedly wags his tail, as he thinks he will get a chance to play with other dogs.

The owner takes the Lab to the other door. The Lab, always weary and fearful of visits to the veterinarian's office, is visibly upset and resists. He thinks he is going to the vet and will have to become territorial once he's in a confined space with the other dogs.

The owner walks each of the dogs through their separate rooms and closes the doors behind them. Five minutes go by. The owner walks over to the room with the Labrador, and lets him out. The Lab growls, barks, and viciously snarls at the owner as soon as he is let out of the room. The Lab is 10 times more upset than when he entered the room and the dog has to be restrained by the owner in order to calm down from the fit that has overtaken him.

Once the owner has calmed his Labrador down, he proceeds to the other door where he left his Retriever. The Retriever comes out running. He wags his tail, jumps for joy, and is absolutely elated as he greets his owner upon his release from the room. The Retriever is 10 times happier than when he entered the room and has to run through the house in order to release all of the joyful energy that has overtaken him.

Clearly, both dogs had vastly different experiences behind the doors.

What could the Lab have seen to make him so upset and angry and what could the Retriever have seen to make him so excited and happy?

What was in the room?

The answer is 'Mirrors'.

Simply, mirrors.

This story is important because it shows that what we often see in the world is a reflection of our attitudes and expectations when we walk through the door and into the room, and not the room (world) itself.

We live in a world of war, terrorism, economic uncertainty, insecurity, failure, shortcoming, and disenfranchisement… and we live in a world of charity, prosperity, wealth, hope, giving, progress, and advancement.

These are all present in the room. So we must choose what we see, what we do, how we react, and how we live.

It is important to always remember that you are in a room full of mirrors. If you choose to see opportunity, then undoubtedly opportunity you will find.

But what if you see hope, opportunity, and positivity in the 'mirror' but know that there is still war, poverty, crimes, crises, in the same 'room' and you want to do something about it?

Good question…

That leads us to our next chapter—to walk through life as The Old Man.

Chapter III. – Tale of The Old Man

This is a tale of an ambitious young man as he walked through life.

This man – like many others around him and before him – wanted to change the world, as all do at one point or another. So this is how the young man approached life… he set out, heart ablaze and mind sharp and keen, to change the world.

The youngster, ready to fight the good fight, traveled to distant lands to right international injustices and to alter the state of foreign governments, for he truly wanted prosperity for those who knew not of its comfort.

This proved to be risky – and after many close encounters with death, the youngster returned home – frustrated, but humbled.

The humbled young man, still ambitious to change the world – shifted his focus from distant lands to his own home nation, because even a great nation can certainly have its faults. So, the humbled young man set out to change his nation… and fight the good fight.

This proved to be taxing – as he spent many years protesting, fighting, arguing, picketing, and debating… much with minimal results.

The fight with his nation took a toll on the man's conviction. By this time, he had married and had settled into a town with his wife and young children.

The man, still ambitious to make a difference, decided to shift his focus once more.

He so desperately wanted the community he was a part of to reflect the ideals he envisioned when trying to change foreign lands, and he became so disgruntled at the lack of results from his fight with the nation – that it became his mission to change the community. So, he set out once more… to fight the good fight.

This fight went well in to the man's elder years. Eventually, his children had fully grown and settled into communities of their own. Still, the man persisted in attempting to alter his own community.
This proved to be challenging, as the community found it difficult to relate to a man who was so disgruntled from his fight with the nation and they were reluctant to share in his ideals. And so, his efforts were in vain and his words went unheard.

It was in this, his darkest hour, that he decided if he could not change foreign lands, his own nation, or the community he would at least try to ensure that his grown children would live a life in accordance to greater worldly principles. Once again, the man set out to fight the good fight… and change his family.

As the man's priorities had been the world around him and not his own family, his adult children felt slighted, forgotten, and unimportant. No matter how hard the man tried, and how well intended his convictions may have been – his family resisted his attempts, as they were content and set in their ways.

At this point, the man was weary from his life of fighting without much to show for it and he retired to his solitude; disheartened, disappointed, and defeated.

Fast forward… at the twilight of his life, the old man was on his deathbed and his family sat around the hospital room he called home. There, they all reminisced on his stories and his lifetime of fighting the good fights.

His grandson, still a young boy, sat in amazement at the stories and asked his own father how to go out into the world, so he could make the changes that were necessary to improve the world around him, as his grandfather had intended to do.

His father replied, "If you wish to change the world – you must first change yourself and become the best version of you. And then you go out into the world and fight the good fight."

It was in that moment that the old man had an epiphany. He realized that he had gone through his life all wrong. The best way to have changed the world was not for him to have gone out and fought wherever a good fight was needed, but to instead, change and improve himself. For if he had walked through life with the principles he wanted to instill in the world, he would have raised his family with the values needed to inspire, collaborate, and lead – to fight the good fight.

His family would have then gone out into the community where take would have taken active roles to make incremental improvements to build a better life. They would have grown to become integral constituents to the progression of the community. Other members, inspired by the actions and generosity of few, would then go on to take actions and strive to build a better community.

The community would then go on to become a transforming, progressing, and bustling microcosm… a microcosm that would become a model for other communities around the nation.

The communities, each transforming and progressing in their own ways, learned to identify problems, work together, take action, and make decisive strides to improvement and advancement. These communities, now familiar with how to learn and grow, would then go on to inspire others around the nation and effectively transform the country.

Thus, the nation would see its influence in greater global society – and then go on to change the world.

So, if you want to change the world… walk through life as The Old Man.

Chapter IV. – The Idiocy that is Racism, The Weakness that is Sexism, The Insecurity that is Classism & Concepts of Tribalism

In our modern society, there is no greater subtle threat to progress than the inherent division between groups of individuals in our nation based upon identifiers that are nonadjustable. The most prevalent of these, of course, are race, gender, orientation, and creed.

These notions of division serve no greater purpose in creating a cohesive, collaborative society. As such, we must recognize the idiocy that is racism, the weakness that is sexism, and the insecurity that is classism. Furthermore, we must also acknowledge the tribalist mentality of separatism and division that are in each preceding occurrence. Each of which inherently divides people into opposing groups or 'tribes'.

The Idiocy That is Racism…

Although we are far removed from the days of Jim Crow and years of Apartheid – racism is still present in every corner of the world. One thing is clear, the denial of a person simply because of the color of his or her skin is essentially the denial of oneself as the brother or sister to a fellow human being.

Racist and xenophobic actions are not only misguided, they are abhorrently shortsighted. The only way to truly progress as a world is to press forward with the advancement of others – regardless of their ethnicity and nationality.

The onus for advancement is based upon systemic foundations, certainly – as systemic causes and influences in the lives of citizens are a major point of contention in our nation. In many cases, this contention can be pointed to the disparity caused by poverty, or a lack of resources, which determine an initial path of an individual's life. Resources, in this instance, can mean stability, education, disposable income, exposure to positive stimulations, etc.

To deny the implicit advantage of having additional resources from birth, through schooling, to early adulthood – and the eventual trajectory a fortunate individual may experience because they have access these resources – is a foolish exercise in feigned ignorance.

However, as luck and skill should have it, "resourcefulness", and not mere "resources", are often necessary to offset those initial negative trajectories inherited by circumstance. It is up to each individual to step up by outperforming any instances of prejudice or doubt by maintaining integrity and dedication to all that may surround them.

We must also demonstrate courage to admit the fallacy perpetrated by likened individuals – regardless of color – and to not identity with others based solely upon race, color, or ethnicity. Instead, we must identify with others through shared values (should they be noble and honorable), as this will allow progress to be the path of least resistance.

This rings true regardless of citizenship or origin. For in this great country, we are all true immigrants – the difference being how many generations removed we are.

There should be no pride taken by a particular ethnic group that maintains prejudicial beliefs. For there are likely an equal number of degenerates represented in every group, as stupidity and ignorance is not solely tied to a single tone of the skin.

To be a great world and society is to champion intellectualism, character, integrity, and the holistic values of inclusion, and to thwart xenophobia and racism at every opportunity. It is of particular importance to combat systemic inequalities by ensuring transparency and opportunities to all.

There is, however, another proponent to equality, which is held by the individual, and it is to not get trapped into the downward spiral of competitive oppression. No one is better when they try to find opportunities to view themselves as more oppressed than another through minor slights or inadvertent actions. Discussing this act alone, in modern society, is for another day.

Put simply, to reject another person or a group of others for mere complexion, tone, and color is to essentially reject another version of one's self.
Hence, the idiocy that is racism.

The Weakness that is Sexism…

Fortunately, for the greater good of the world, women in developed and developing nations are headed in an upward positive direction – as both women and men have access to similar education, resources, and can be viewed as potential leaders to create substantial change in the nation and the world. All of these are fundamental to respect, acceptance, and valuing the contributions from those of opposing genders.

Combine this egalitarian advancement with the inherent need for empathetic understanding regarding the various challenges people face in everyday life – regardless of gender and social setting –and similarities will be found where differences were once common thought.

A major component in inequality that women currently experience involves another gap of the economic variety – pay. Primary research indicates that women earn eighty percent to their male equivalents in professional environments.

There is not, however, enough information in this figure to generate any actionable changes to policy, practices, or career trajectory. Additional research to further segment this data would lead to a more certifiable conclusion to determine the cause of this discrepancy in compensation. Such research may suggest a clear course of correction or advisement – whether it is difference in industry, education, trade, certification, role, recruiting/negotiation, regional market, year's experience, or any number of major components.

The U.S. and other developed nations must create the means to accurately assess the landscape and the various factors that play into the compensatory decision-making of hiring managers and executives. This kind of assessment can be accomplished through a federal, private, or quasi-partnered approach. But to begin vying for a solution without an accurate portrayal of the total landscape is to attempt to throw darts at a board in total darkness – an exercise that will lead to far more misses and accidents than the incremental improvement necessary.

The street, in this regard, goes both ways though – as societal expectations of masculinity and femininity are strong causes of insecurity, frustrations, and sometimes of attack. Much to everyone's annoyance, these expectations then lead to the cumbersome phrases that begin with statements like, *"A man/woman should…"* or, *"A real man/woman would never…"* etc.

These imperatives are, of course, utterly hollow and offer minimal merit to accuracy. They do not take into account the individuality that comes along with one's own personality, traits, characteristics, and interests – as these are independent of gender.

Instances of personal attack or aggressiveness are not subject to women alone, and any belief that men and boys do not experience pressures of their own to conform, fit in, and align to groupthink against their own true north is also blind. Again, no one is better for comparing hardships they come upon based on their nonadjustable categorizations.

The message is that anyone can experience challenges or biasness from their gender regardless of the gender of the perpetrators. To have combatant beliefs and attitudes towards others simply because of a difference in chromosomes, and the physiological characteristics that follow, is completely asinine and a weak point of argument.
Hence, the weakness that is sexism.

The Insecurity That is Classism…

For the greater part of the past century, there has been a growing disparity of wealth amongst C-suite executives and business creators, managing partners, salaried supporting positions, and wage earners who build and service various components that contribute to the successful operations of a company or organization.

This division goes well beyond the simple economics of wealth generation and earning power, as a separatist mentality is present amongst differing individuals in each positioned category.

Classism, in essence, involves the viewing of another group of individuals negatively because of what they have, want, value, or admire.

From constituents to politicians…

From associates to executives…

From landscapers to real estate moguls…

From those of small town humility to jet setting travelers…

There exists a wide range of people who hold various values, concerns, fears, and dreams.

Using Maslow's hierarchy of needs as a base point for understanding the wants and needs of a person – then food/shelter, safety/security, belongingness, self-importance/value, and actualizing one's purpose are what are most often sought after. This is true no matter one's background, or where a person comes from, or if he were a prince, a pauper, or somewhere in between.

From shotgun homes, to the projects, to suburban homes or urban high-rises – all are forms of homes that provide a space for living. Obviously, each form of housing is different and each embodies a different connotation of opulence, or lack thereof. But, is one innately better than another? After all, beauty, desire, and ambition are in the eye of the beholder. The short answer is, "no". It's just that one group has more resources and influence than the other – which may leave the other groups with a contentious sense of disapproval.

A division in value seen within people of different segments of society undermines the importance of understanding and appreciation of the value given by another segment. After all, an organization or government entity with all executive leaders and no group to implement strategy or policy is not a successful one – and one that is filled with groups of only implementers will be left without direction.

This given value by separate groups is required for the interdependence we have and will always need in order to operate a successful and progressing society.

Classist mentality can be relegated to any individual's innate need to exhibit a sense of superiority from a lifestyle that is either inherited or afforded through good fortunate or personal shortcoming. An individual, given enough time and effort, could experience either, which points to a subconscious inferiority thought-process that the individual may seek to overcome.
Hence… the insecurity that is classism.

Concepts of Tribalism…

I hope you have noticed the dichotomous "us versus them" mentality that has been present in every preceding point in this chapter. The "us versus them" thought processes do not lead to a greater life nor do they lead to a greater understanding of one another. For if you focus on an overarching category, you miss the forest through the trees and fail to realize that the divisive categorization of subgroups aids in the advancement or progress of no one.

Why do we do this to ourselves?

Is it our upbringing?
Is it learned behavior?
Is it the media?
Is it patterns of thought that have turned cynical?

Who knows for certain? Blame is not what is important in these instances. Instead, it is acceptance, understanding, camaraderie, and an admiration for the life of another that leads to a greater understanding of your fellow man. And that is the intention in rejecting tribalist notions. We must see each other for who we are – and not merely what we are.

Let us no longer seek differences in one another, but realize that we are far more alike than we are different. This will enable us to truly see the idiocy that is racism, the weakness that is sexism, and the insecurity that is classism.

Chapter V. – Reeducating Education

Education has long been viewed as the great potential equalizer for achieving social mobility in America. In pursuit of the American Dream, the story goes that one studies hard, masters their education year after year to gain a basic understanding of subjects of importance, and continues on a path of higher education to further specialize in an area of interest. Upon graduation, individuals are then set on a trajectory toward financial security and they can expect to receive steady income for the longevity of their careers, which will afford them a life of comfort and prosperity.

However, the far more untold story is the educational race to the middle of mediocrity; it is seldom class mobility. Instead, individuals are relegated to lives that are dependent upon performing dissatisfactory duties to maintain a meager lifestyle. Worse yet, are the most common instances in which education thoroughly fails to prepare individuals for the harsh realities they are bound to experience – absent the proper perspective.

It is time that we look to reeducate education.

Educational institutions through high school still mimic the educational processes from the dawn of industrialization. It can be argued that the purpose of this education style was to educate class upon class of factory workers with the skills needed to just take direction, learn repetitive actions, and replicate those actions with accuracy and with minimal instruction and oversight.

Programs today lack resources to create more flexible educational approaches that are needed to create environments for individuals to explore an interest, experiment with it, understand it, and achieve the mastery of it. The aptitudes needed for advancement in today's competitive economic landscape requires a finesse of social skills that are often untaught, but are instead picked up indirectly through secondary and tertiary means.

While individuals diverge to different academic or professional paths following high school, it is the soft skills, and basic life teachings that are paramount to achieving success in work, home, and life. People benefit from learning skills such as public speaking, how to conduct oneself in an interview, and how to convey their ideas to others.

Developing a sense of self-awareness and self-assessment and the ability to participate in critical thought and analysis, debates, and socially conscientious conversations contributes to a lifetime of value and personal insight. Unfortunately, critical lessons about personal finances and lifestyle budgeting, basics of intelligent consumerism, investments, retirement, insurance, credit, mortgage, financing, as well as healthy living, wellness, and happiness are not universally taught, as they should be. Nor are they stressed with the pertinence they so rightly deserve.

If asked, most people would agree that these are topics worth learning and are, in fact, absolutely necessary in the modern world. Yet, the majority of curriculums do not focus heavily on these subjects.

In order to create opportunities to teach and learn these skills, monumental shifts in our educational culture are necessary. Educators need to be given proper power and flexibility so that they can introduce these crucial soft skill subjects, and not be solely limited to teaching subject matter that contain metrics of standardization for evaluation – which, at its core, creates indistinguishable academic graduates.

Suffice it to say, the research, suggestions, and implementation of these shifts would require its own work to appropriately address the methods of shifting the culture.

Two major components of education where shifts are also necessary include:
     1.) Reducing the stigmatization of non-academic paths and creating transparency to comparable opportunities.
     2.) Minimizing wealth disparity of the educated classes.

A large proportion of primary education institutions articulate that only way to success is through continued higher education – and the pursuit of college/university studies and advanced degrees. While in many cases this is true, there are an equal number of cases in which a student may be better suited, have greater aptitude, for labored and skilled positions. Instead of stigmatizing and culturally lessening the impact that blue collar and labored positions hold, we must do a better job of championing and creating apprenticeship programs and paths to traditional mastery. Currently, students are often not advised of these types of opportunities until they forfeit the path of higher education.

Secondly, for those who elect the route of higher education, upon completion of their degrees, they will be entering a job market that is more competitive than ever before. Furthermore, they will be competing directly with other similarly educated individuals with whom they share a minimal true differentiation in skill. This dynamic effectively dilutes the value of higher education and has the dual dagger effect of burdening the less fortunate (i.e. students from less financially-able families) with debt that inhibits near-term financial security. This trend will undoubtedly have much greater macroeconomic effects as the decades go on.

This situation cannot be sustained indefinitely. Policy makers, academic institutions, educators, and the educated must be forward thinking concerning the markets of the future to determine the needs of our citizens through educational systems. We must work together to reeducate education.

Chapter VI. – Concepts of Climate Change

Misnomer (noun); a wrong or inaccurate name or designation.

In this chapter, there will be several principles outlined that tie into one of the greatest single threats to the longevity of mankind – climate change. At the end of this chapter, a new term will be proposed altogether – one that is more apt to categorizing human activity that contributes to the outcome of climate change.

First – a series of background points to keep in mind as you read on.

1. We have a tendency to pick several words that become a phrase we use describe an event or occurrence. The phrase is then often attributed to one individual who has "coined the term." Think of debt ceiling, financial cliff, great recession, etc…
2. Initial predictions and reactions to theories and ideas are not always entirely accurate. Oftentimes, the public remains uneducated as to the potential impact of an idea, whether it fails or leads to success. Regardless, it is common for individuals to take a spectator's point of view. Take, for example, the development of powered flight and how some timely predictions proved to be completely inaccurate and foolish.
3. People tend to remember stories far better than facts and figures. Much of human history has been recorded through stories that have been told generation to generation through the ages. This chapter

continues with a few simple stories correlated to climate change/global warming, as well as introduces an additional term to describe the shift in climate due to man made activities.

We begin with a story about Samuel Pierpoint Langley.

Mr. Langley was a pioneer and early mechanician of human powered flight. Langley's plane, known as the Aerodome, made two attempts at sustained flight in October and December of 1903. Both flights were unsuccessful in allowing pilot, Charles Manly, to take to the sky – rendering both attempts as complete failures.

Shortly after Mr. Langley's first failure, The New York Times released an editorial titled, "Flying Machines Which Do Not Fly".

Those who penned the editorial all but mocked Langely and his dreams of powered flight. They even confidently proclaimed that it would take, "one million to ten-million years", for mathematicians and mechanicians to create sustained flight.

On December 17, 1903, just nine weeks after the publication, the Wright brothers flew their first flight in Kitty Hawk, North Carolina – for a grand total of fifty-nine seconds. A short sixty-six years later, man walked on the moon. And now, merely over a century since Kitty Hawk, private innovators are once again advancing mankind, as we are on the brink of establishing human footsteps on Mars.

Clearly, the editors were sorely mistaken, and their hyperbolized viewpoints were far from accurate.

Take this inaccurate initial prediction by the editors of The New York Times into consideration with respect to modern debates over climate change and global warming. No, it is not a hoax that is made up for the sake of dramatization. It is real, it is observable, it is explainable, and it is predictable.

Mankind has invented machines, pioneered industries, and mastered the land, the sea, and the sky. All of these accomplishments were achieved on purpose. It is foolish, shortsighted, and moronic not to see that we are also capable of incidental destruction. Any denial of the human influence on rising global temperatures through man-made industry and processes is a pure mental fallacy; and it showcases a lack in understanding of the scientific principles at play.

Climate change/global warming must be observed where the initial impact is most significant – through the thermoregulatory influence held near the polar icecaps.

The interesting fact about the polar ice caps – and ice in general – is that H_2O in a solid state (frozen) is less dense than its liquid state. This is because of the molecular structure and the oppositional forces of the hydrogen bonds – effectively forming what is called a "bent" angle.

DIAGRAM OF H₂O MOLECULE

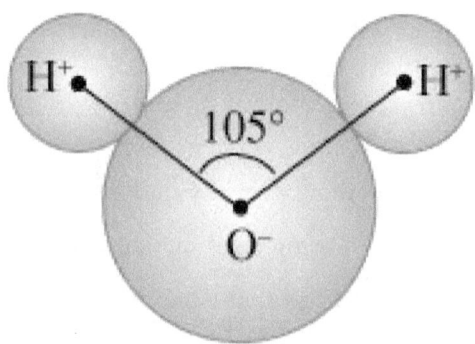

As water transforms into ice, it creates interconnected crystalline structures and it displaces more surface area and volume than before, thereby creating a buoyant mechanism.

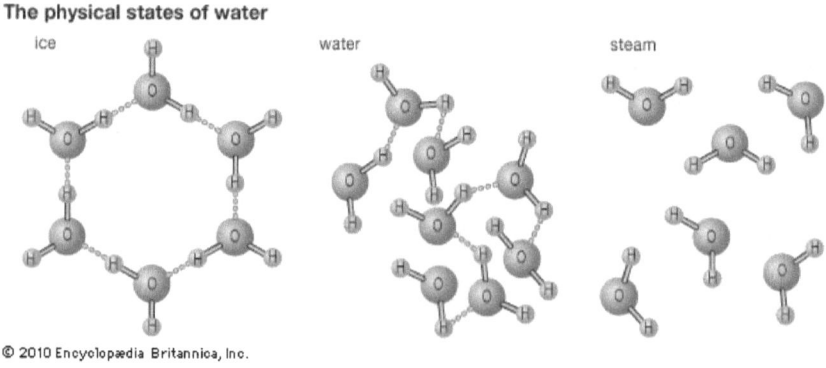

This mechanism also means that the thermal properties of ice influences both air above and water below.

As it stands today, we are rapidly losing the northern ice cap – and it will disappear altogether, if we do nothing. This holds true for the southern ice cap as well.

To show the implications of how the ice caps contribute to our environmental stability though thermoregulation, here is an analogous story.

When I was 19, I was a sophomore transfer to LSU. Like many Americans today, I was a self-funded student. Being self-funded meant that I needed to work to afford my tuition, books, rent, and food. However, I did not have a vehicle at the time. So, I took the $1,500 I had in my savings and purchased a blue 1994 Jeep Cherokee – as is, meaning I had no warranty should anything go wrong with the vehicle.

This Jeep had a broken air conditioning system. The compartment that held the Freon (the cooling agent in our AC units) was split in half and the tubing was rotted. I could not afford to replace everything immediately. But this was okay, because the 100° F Louisiana heat was bearable enough to travel between my class and my work.

Some months went by, and eventually the coolant reservoir, which held the engine coolant, cracked. Due to the antiquated detection systems the vehicle, this situation went unnoticed until the coolant eventually ran dry. Without coolant, the engine overheated, locked up, and ceased to run. It cost more to fix the engine than to replace the old vehicle, effectively making the Jeep completely totaled.

Fortunately, by this time, I had established enough of an employment and income history to finance a quality used car. I purchased a 2009 Nissan Maxima, which came with a warranty and service package. This meant that if the Freon ever leaked, or the engine coolant ran out and the detection system failed, and the engine overheated and locked, that the warranty would cover damages and replace the engine, parts, or the entire car. No matter what happened, I'd still be able to drive to work, make money, and afford my tuition, books, rent, and food for the year.

Now... think of the ice caps as Earth's natural Freon and engine coolant. Its thermal nature significantly cools all air and water that comes near it. Combine this with the Coriolis effect from the Earth's rotation (whereby a mass moving in a rotating system experiences a force acting perpendicular to the direction of motion and to the axis of rotation) and the result is the trade winds that act as the planet's "air conditioning". This effect also generates a "coolant system" from the water currents it creates. Both of these combine in maintaining the climate, as we know it today.

CORIOLIS EFFECT DIAGRAM

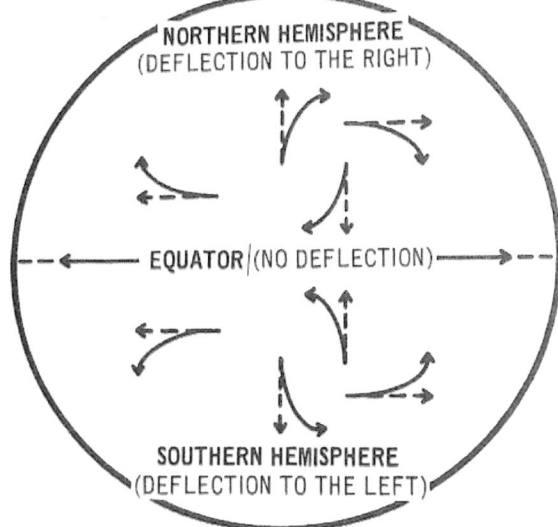

TRADE WINDS DIAGRAM

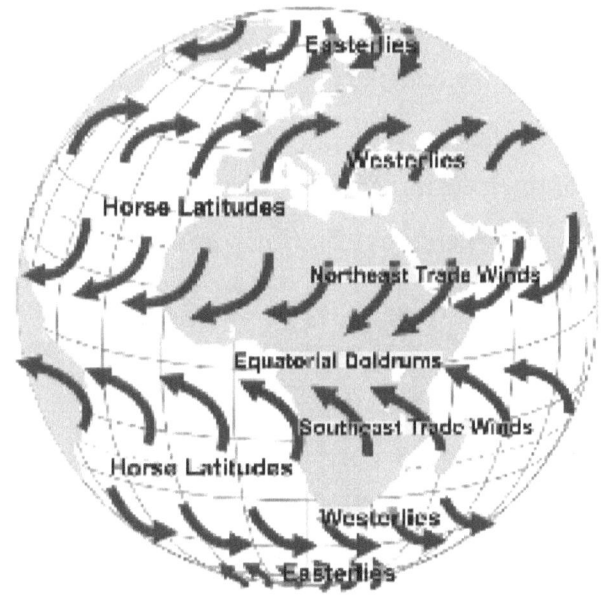

WATER CURRENT DIAGRAM

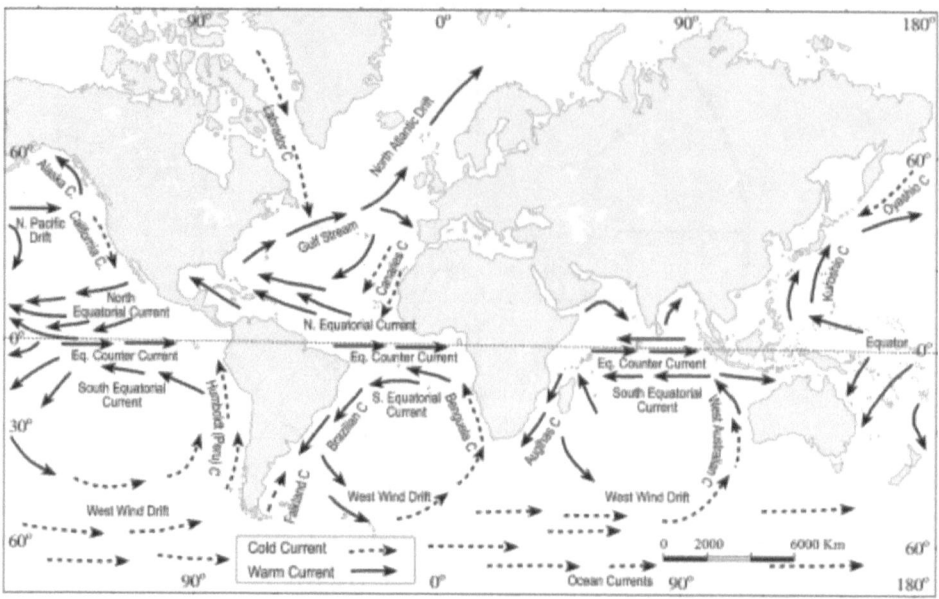

Remove the Freon and the engine coolant – what happens?

Unlike the story of my Jeep and Maxima, we don't get to finance a new Earth on credit. We get one planet to live on (for now at least).

We cannot deny the loss of the ice caps. We must take action to minimize the effects we have on our environment. And, with it, we must understand several concerning rates that are observable in our modern day and how they contribute to what I describe as "Foundational Climatologic Destruction", or F.C.D.

Chapter VII. – Foundational Climatologic Destruction (F.C.D.) & Rates of Concern

Foundational Climatologic Destruction (Noun); Reoccurring actions of human behavior that severely disrupt, inhibit, or destroy the ecological systems that life depends on – whether incidentally or intentionally at scale.

The term Foundational Climatologic Destruction is being introduced to explain the root (or foundational) causes of climate change and why it is imperative that we focus our effort on minimizing, stopping, and reversing this destruction.

To better explain the premise for the term, imagine, for a moment, a skyscraper that towers over a city filled with businesses and shops. Notice how it stands powerfully as compared to the smaller buildings that surround it. Now imagine a bustling apartment complex, filled with all kinds of tenants. Now think of a house within a small neighborhood.

A skyscraper, an apartment complex, and a house have vastly different needs (materials, labor, planning, maintenance, financing, etc.) while being built. But the first thing needed in building each structure is a strong foundation that it can stand upon. Destroy the foundation and it won't matter what was you used to build the building; it will eventually collapse and fall to the ground.

We are, in essence, eroding the foundation that has enabled the thermoregulatory environment that has enabled human life (and other forms of life we depend on) to evolve and thrive.

We must work **together** to fix, alter, and replace the systems that cause F.C.D.

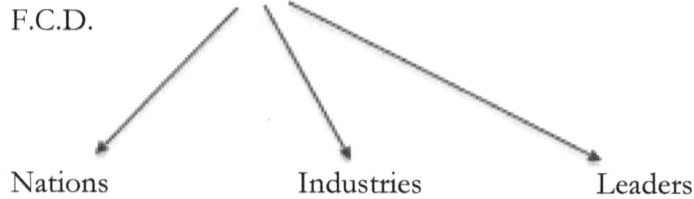

Nations   Industries   Leaders

To better understand some of the key components that lead to climate change, here is a brief list of some the current culprits contributing to F.C.D, as well as concerning rates of human, industrial, and economic growth that will likely increase these key components if we do not intervene.

Concerning Rates:
- Growth of the global population
- Growth of global energy demand
- Growing rate of carbon emissions
- Growing rate of emissions of other complex gaseous compounds
- Rate of deforestation
- Rate of ocean acidification
- Rate of ice loss in polar ice caps
- Increased auto production
- Increased global travel via plane
- Increased trade via marine transit
- Increased animal agriculture and rates of methane emissions

The following charts and mathematical examples exemplify the rates listed above:

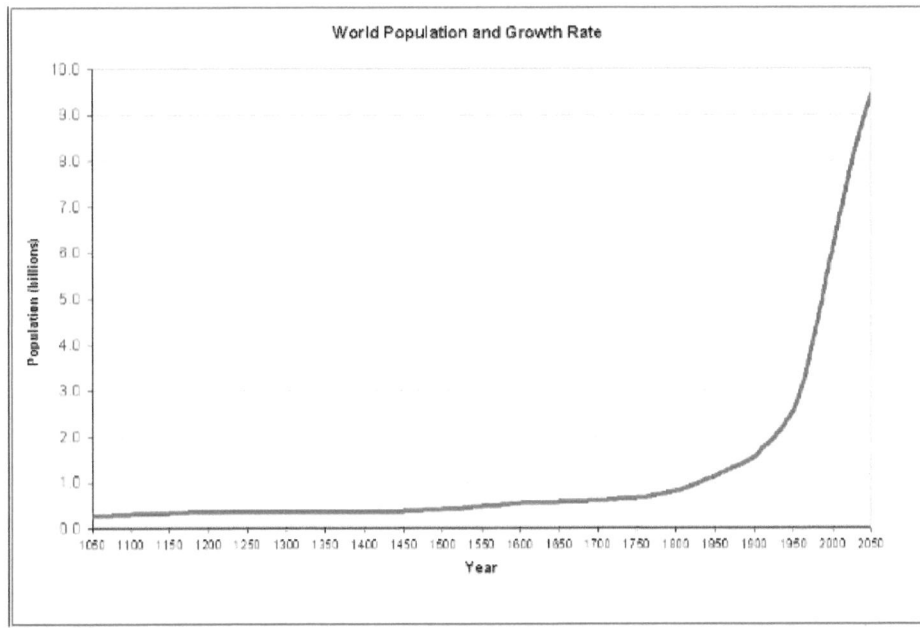

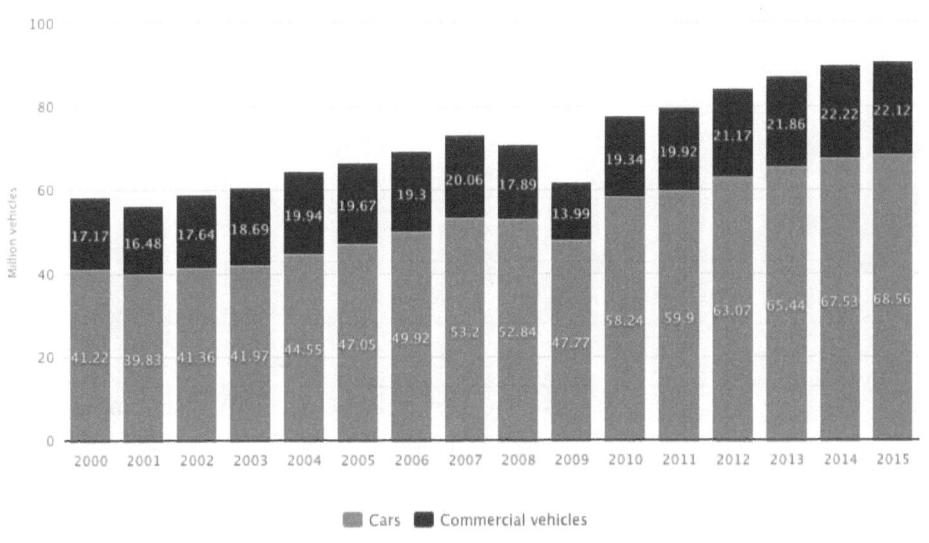

© Statista 2017

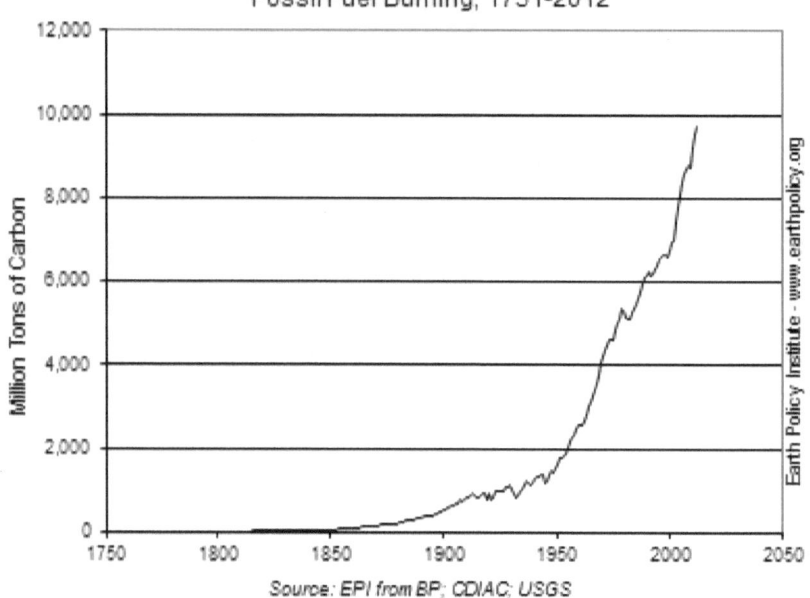

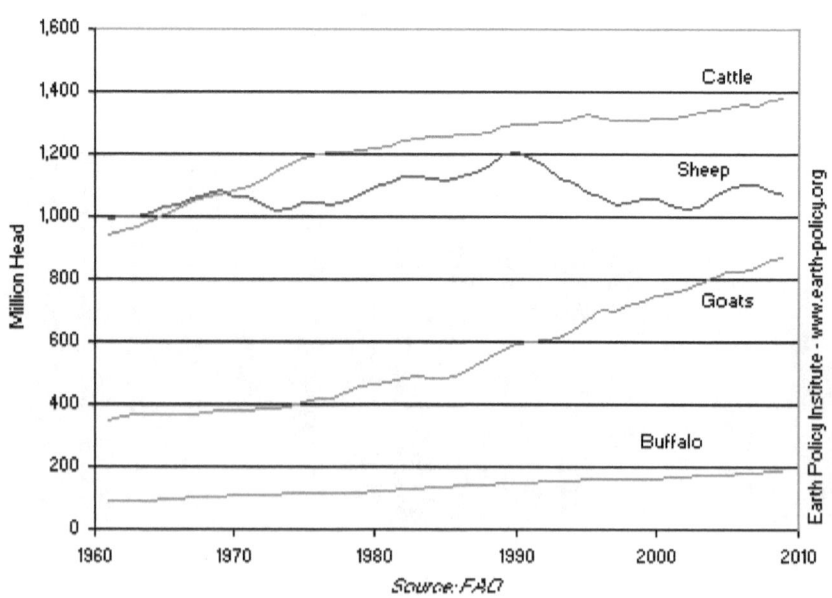

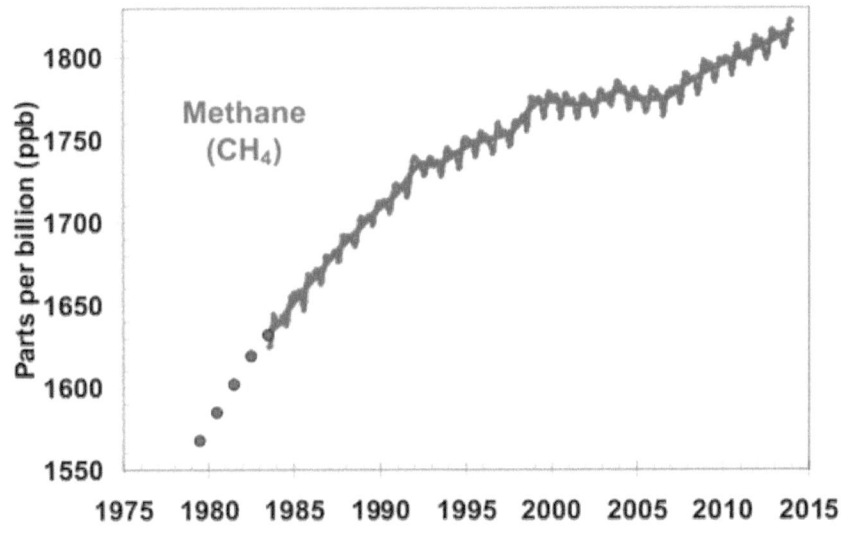
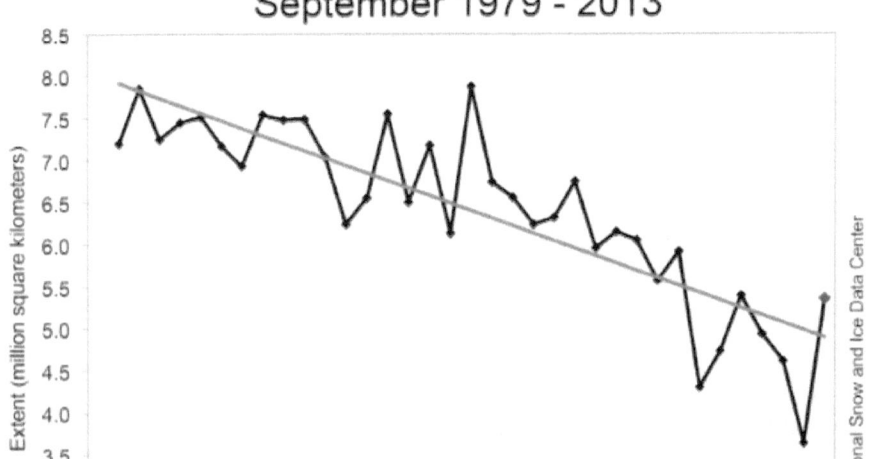

Functions associated with concerning rates:
- Increased thermal absorption/retention in atmosphere due to increase in complex gaseous compounds
- Increased temperature at ice caps
- Loss of ice caps and the reflective nature of the white ice
- Decreased reflection of photon (light) energy from the sun and due to less white ice
- Increased absorption of photon (light) energy from the sun and absorption of thermal properties in additional dark water caused by less white ice
- Increased rate of ice loss exacerbated by increasing rate of thermal retention of water surrounding ice caps
- Decreased protective specific heat of ice caps due to ice mass loss which then increases susceptibility to additional rapid melting

Perhaps the best way to showcase Foundational Climatologic Destruction is to view the sequenced chain reaction of events and their effects/influence on the factors that follow.

The easiest way to visualize this principle is to utilize the algebraic viewpoint of the function of a variable, aptly viewed as the function of variable x equaling the output y; or $f(x) = y$.

Then, viewing the function of y equaling the output z; or $f(y) = z$. Then, adding the entire summation of all z's outputs to view the cumulative effects these factors have that contribute to climate change; or $\sum z_1 + z_2 + z_3 + z_4 \ldots$

Example....
(Abstract views)

x = Auto transit with the internal combustion engine

$f(x) = y$ where y equals an increase in the release of complex gaseous compounds (carbon dioxide)

$f(y) = z$ where z equals increased thermal retention in the atmosphere

x = cattle livestock that feeds the U.S. population

$f(x) = y$ where y equals an increase in the release of complex gaseous compounds (methane)

$f(y) = z$ where z equals increased thermal retention in the atmosphere

Example....
(Quantifiable view)

x = automobiles driven in the U.S. in a single year

$f(x)=y$ where y equals CO_2 emissions released into the atmosphere per automobile and is multiplied to total CO_2 emissions released by all cars

$f(y)=z$ where z equals estimate of increased thermal retention per volume area of CO_2 emission release and multiplied to total

Example in action...
(Quantifiable view)

x= number of vehicles on U.S. roads in 2016

$f(x)=y$ where y equals total metric tons of CO_2 released from vehicles annually

$f(y)=z$ where z equals the quantifiable thermal retention of (1) one kg of CO_2 multiplied by total CO_2 released

---

x= **253,000,000 vehicles** on U.S. roads in 2016

(Note: 4.7 metric tons (4,700 kg) of CO_2 is released from 1 typical passenger vehicle per year)

$f(256,000,000 \text{ vehicles})= y \rightarrow$ 253,600,00 vehicles on the road in the U.S. in 2016 multiplied by 4,700 kg's of CO_2 released from 1 typical passenger vehicle per year

= **1.19 trillion kg's of CO_2 released** in the U.S. in a single year due to automobiles

$f(1.19 \text{ trillion kg's of } CO_2 \text{ released in U.S. annually})=z \rightarrow$ thermal retention of (1) one kg of CO_2 multiplied by the 1,190,000,000,000 kg's of CO_2 released into the atmosphere by automobiles driven on the road in the U.S.

= **Total thermal retention increase PER YEAR.**

Bringing it together...

We must do calculations and calibrations for every man-made process or human activity that contributes to climate change, i.e. any function that creates and releases complex gaseous compounds in the environment that results in an increase in thermal retention of solar energy.

Here are a few examples of man-made and human activity what we must calibrate, study, and create sustainable solutions for...

- **Transit** – internal combustion engine with gasoline and diesel fuel sources; auto, marine, aeronautic, etc.
- **Animal agriculture** – methane producing food sources; livestock.
- **Manufacturing** – production methods that create complex gaseous compound bi-products without 100% filtering or repurposing bi-product creations.
- **Energy creation** – coal, natural gas, and fossil fuels; electricity and utilities.

We know the trends that result from climate change; increasing air and sea temperatures, loss of polar ice caps, increasing sea level, harsher weather patterns and natural disasters, and, ultimately, permanent alterations to, and the loss of, highly populated land masses – not to mention mass extinctions.

It is possible to work backwards to identify every factor that adds to the concerning rates and components within Foundational Climatologic Destruction. We must work together to determine its level of influence and contribution to the overall threat, and we must curate solutions that either significantly reduce or entirely eliminate said factors. This is true across the world. As such, we need to develop the means to create sustainable solutions that enable the economic inclusion and support of former systems – while shifting to economically and ecologically beneficial means.

The costs, commitments, and compromises to doing so would be significant and will not occur quickly. But, it must be viewed as a long-term investment to ensure our sustainable future. And, with it, we must maintain a diligent resolve to create a better tomorrow, so that we have a world that is worth passing on to future generations.

"A society grows great when men plant trees whose shade they will never enjoy."
    - Ancient Chinese Proverb

Chapter VIII. – Ideas of Ages Past

The building of our nation over the past couple hundred years and our world over the past several millennia has yielded tremendous knowledge – from cave paintings, art, harnessing of the power of fire, to modern advances in technology, agriculture, and our scientific understanding. Our achievements have allowed man to master the land, the sea, and the skies. We are well on our way of mastering the stars. And one day, we will hopefully master the land, sea, and skies of foreign celestial bodies.

Of course, the road to all of this mastery has not been one of complete merit. Throughout history, man has used methodologies and held notions that have not been aligned to a true north of a moral compass. From slavery to serfdom, indentured servants to child workers, imperialistic exploit and indigenous persecution, to religious rejection of scientific principles, these beliefs have stifled innovation and thought.

One such fallacy was the geocentric belief that placed Earth at the center of the universe and labeled the accurate, heliocentric view, in which the sun is the center of our solar system – as heresy, punishable by torture and death. Although, some may still believe man to be the center of the universe – or much worse yet, they themselves to be the center – obviously, this is not true. This centric view of position within society is also a common theme in other ideologies. So, let's explore the fallacy within ideas of ages past.

The White Man's Burden

The White Man's Burden is an archaic belief, which dictated that the Caucasian race needed to teach other races of 'lesser' ability how to thrive in a modern world. This belief can be traced back to a poem written by Rudyard Kipling in 1899, during the Philippine-American War.

Kipling, like many white Americans, urged others to embark upon pursuits similar to that of an empire. The thought, The White Man's Burden, was then used and portrayed as the Caucasian man's moral and sacred obligation to rule and preside over colored and non-white populations – be they American or foreign.

Through this belief, it was urged that the burden included overseeing the economic, cultural, and social progress of others, through colonization and imperialist exploits until such a time that the non-white populous had actualized their independence and could handle their own personal affairs.

This thought process effectively created an inferior view of non-whites by Caucasians, and became the basis for many systemic beliefs of supremacy and laws that would frame this mindset for decades to come.

Manifest Destiny.

Manifest Destiny was held as the common thought during the 19th century. Manifest Destiny dictated that the U.S. expansion across the remainder of North American was inevitable and it challenged the average man to rise to the occasion, pursue this vision, and play his part in this certain destiny.

The overarching theme prompted men to command global exploits, conquest indigenous populations, and expand and lay claim to unclaimed land as their own.

The most common ideas presented within Manifest Destiny were the unwavering value of American virtues and institutions within her borders, the inherent mission to spread these institutions and actualize a model world image to that of the United States, and to ensure a religious destiny meant for all by the hands of God. When viewed within this light, the mission of Manifest Destiny can be seen as having been a positive pursuit worthy of the American people.

However, one must also take into account the populations that were displaced during this rapid expansion of land and the consequences from conquests of western territories that eventually were entirely annexed and acquired. During the aggressive expansion of U.S. territory in the 19th century, Native American populations were misplaced, mistreated, abused, and ousted. Genocidal persecution prevailed in order to make way for American expansion.

Manifest Destiny, like The White Man's Burden boils down to another instance of superiority – this time, of the Anglo-Saxon variety with the religious undertone of the movement.

These beliefs were viewed as being essential and were likely responsible for the rapid development seen in centuries past. Manifest Destiny provided the economic strength and natural resources to eventually lay the path of continued transformation for the U.S. to become a superpower. However, these beliefs are no longer needed. They are neither relevant, nor useful, and they are truly ideas of ages past.

A more holistic, modern, and inclusive set of ideals are required now. And with it, lies a needed attitude of succession and ambition for the prosperity for others. There must also be an equitable force to shift economic ideologically from accumulating spoils for personal gain, and instead passing these spoils to following generations of the outstretched, eager hands that are seeking aid and knowledge.

These ideals and motivations must point to those in society who can be labeled as 'the accomplished' and 'the achieved.'

Perhaps the easiest rule of thumb is this: if the universe should deem you worthy in your personal endeavors and give you prosperity and good, then it is your obligation to return good into the world at least tenfold.

"Good", in this case can mean a number of things – wealth, accolades, fame, and prosperity, and most importantly, the inherent lessons that have been taught by the world. It is often the lessons that life teaches us that are the greatest pieces of knowledge we can possess. Lessons from the world are not always easily taught, for the world is designed to tear people down and break us.

The world breaks each of us in our own way, in our own time, and manifests itself differently. The responsibility to mend these breaks caused by the world rests upon the individual. Those who have been broken have two obvious choices – to grow bitter and angry at the world for having broken them… or acceptance. Understandably, if one should choose the first, they may then go into the world with malicious intent and seek opportunities to break others out of spite.

This is not the world's intent, nor why it was designed to break us. Instead, when the world breaks us and we are mended, we are then entrusted with a single sacred duty. And the duty is this; to go out into the world, find areas that are so desperately broken, and stop at nothing to mend them. For only those who have truly been broken know best how to tend to the world.

This… is the opportunity of the achieved, the responsibility of the accomplished.

Chapter IX. – Shoulders of Giants

As I sit here writing, I can't help but notice the lights in my room. I think about how far technology has come, and how I do not need the assistance of candlelight to continue my work well into the night. I see the light in my lamp, and I think of Thomas Edison and Joseph Wilson Swan. I see my laptop and think of Bill Gates. I see my phone and think of Steve Jobs. I see my wallet and credit cards and think of J.P. Morgan. I see my car keys and think of Henry Ford. I see a Tesla across the parking lot and think of Elon Musk. Then I think of SpaceX and from there I think of NASA, Apollo, and I think of Buzz Aldrin and Neil Armstrong. I think about these men and see Einstein, Galileo, Oppenheimer, Nikola Tesla and hundreds of others alike.

And it becomes incredibly clear that we – as a global society – stand on the shoulders of giants. As they have stood on the shoulders of giants before them, and they, on the shoulders of giants before them – generation after generation, century after century, millennia after millennia.

But what I have yet to hear is why we stand on the shoulders of these giants…
And I believe that we all must, in our own way, grow and change – so that we may have our shoulders grow so great that future generations can stand upon them.

But I fear that as a world, we've forgotten that giants – although large – may still fall.

Which brings us to the final chapter of this book... The Purpose of Life.

Chapter X. – The Purpose of Life

"What is the purpose of life?"

This is a question pondered by many since the dawn of man and it is often a source of contentious debate in our struggle for self.

This question has typically been asked with the assumptive, "my", in front of "life", in that we often try to create a perception of life around us and purpose in relation to our own individual existence.
Furthermore, the question of the purpose of life has been asked from the perspective that people – humans, Homo sapiens – supersede other forms of life are not in the same overarching category as animal life, or plant life, or bacterial life.

This view is far from objective. I will challenge you to look from an alternative perspective, removing yourself from life and your place within it.

If you look into the still waters of a pond, or glance out to the ocean, wander across a plain, a desert, or tundra, or climb up a mountain – you will see dynamic environments with all forms, shapes, sizes, characteristics, and behaviors that are life. You will also see yourself included within these environments.

So what is its purpose – life?

I propose, simply, to be.

And I will argue with that, its purpose, in essence, is to thrive.

To grow.
To change.
To challenge.
To innovate.
To survive.

To press forward for the sake of living, because it is life itself that is its own purpose – to live.

And this is the gift the world has given us – to live our lives with purpose.

A common question… is it evolution or intelligent design?

Why not one and the same?

After all, a civil engineer designing a bridge would have to take into consideration an abundance of variables that the bridge will likely experience. The engineer must design the bridge so that it can adapt to handle temperature variations, weather conditions, all while maintaining the function of allowing traffic to cross and maintain structural integrity.

The bridge was designed intelligently – with intelligence behind it – to adapt, survive, and thrive.

Life is no different; it must adapt, so that it may survive and ultimately thrive. But life is far more dynamic than a bridge. Sir Charles Darwin knew this to be true from his research efforts in the Galapagos. Here he observed physical variation in avian species and discerned that individuals within a species that adapted the best and the quickest would eventually survive and thrive.

It is important to note that Darwin said it is the survival of the "most adaptable" and not the "most fittest" when describing which would survive and thrive.

So that is life's purpose… to adapt, and survive, so that the living may thrive.

With that, challenge yourself…
Grow.
Change.
Challenge.
Learn.
Innovate.
Improve.
Press forward for the sake of living so that you too will adapt, survive, and thrive.

So that one day… the universe may too learn to know –
The Influence of Man.

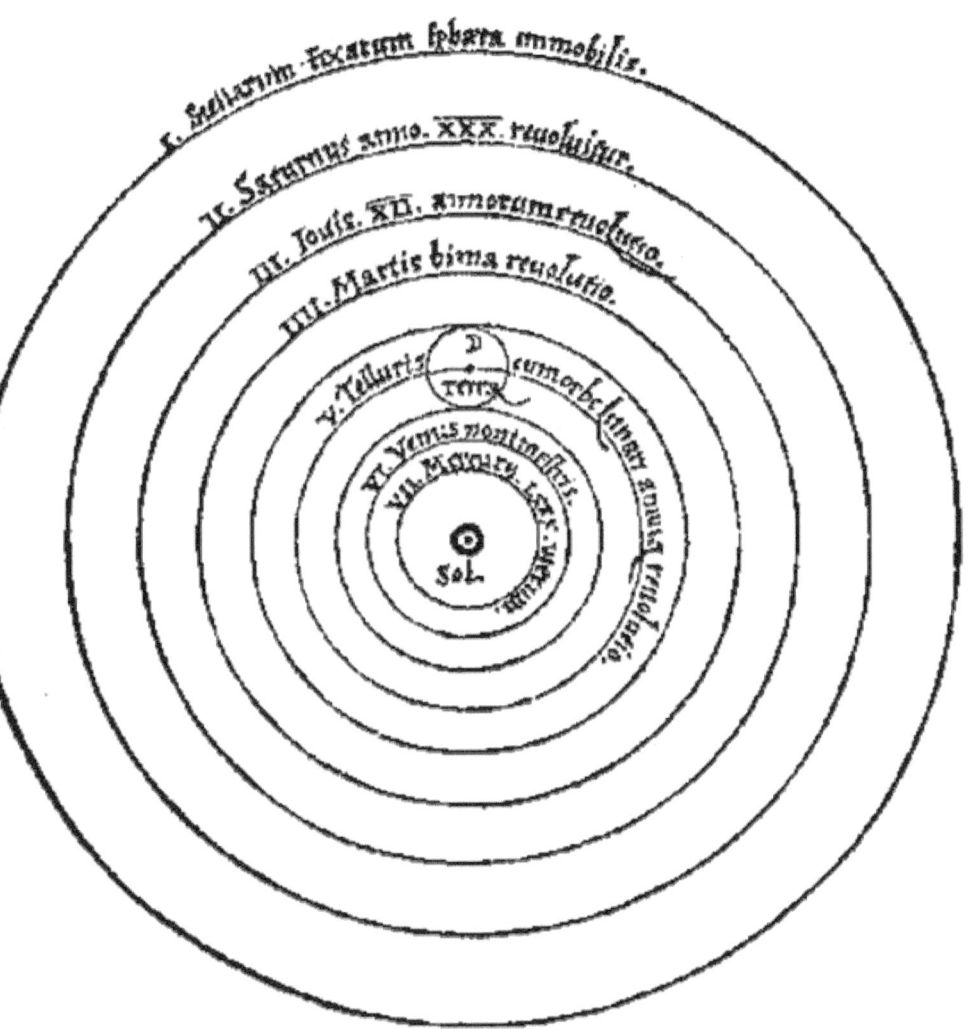

Dedications

I believe it is best to save dedications for last – after all, its best to have an audience follow along when reading, and hopefully come to a similar conclusion in order to understand the impact individuals have made on the author behind the book.

First and foremost, to my mother – thank you so much for everything you have given me. Your selfless and unconditional love and support has been the bedrock of my confidence to go forward in life. Thank you for letting me grow and learn in my own time and in my own way to walk in the world.

To my friends – thank you for the good times and the challenging conversations. I've been very fortunate to find some incredibly intelligent, ambitious, and gracious friends on this journey of life thus far – this is true from Shreveport, to LSU and FIJI, to Gartner, Facebook, and beyond. I've learned more through our interactions and collective successes and failures than I ever did in the classroom. Thank you for humoring me in my long discussions of purpose, business ambitions, and plans for a better world – particularly while inebriated.

That brings me to something I never truly thought I'd say as an agnostic – thank you to a new faith in some form of higher power. Whether your name is God, Yahweh, Allah, Elohim – or if you have no name at all – your presence in my life is difficult to ignore. Thank you for guiding me through my life to this point. Your subtle hints often went ignored, and now I'm hard-pressed to deny your underlying Presence. I will do my best to use the gifts given to me for the betterment of the world, and I hope that with this others may, too, use their unique gifts for the advancement of mankind.

Just know that I resonate most strongly with the words of Marcus Aurelius...

"Live a good life. If there are gods and they are just, then they will not care how devout you have been, but will welcome you based on the virtues you have lived by. If there are gods, but unjust, then you should not want to worship them. If there are no gods, then you will be gone, but you will have lived a noble life that will live on in the memories of your loved ones."

Lastly, and most importantly, I dedicate this book to you – the reader. Whether I know you or not, whether I'll ever meet you or not, whether the world will know your name or not – I hope this perspectives discussed in *The Influence of Man* can help guide you as they have guided me.

Thank you for taking the time to read. And remember...

We are each the protagonists in our own story, and we are all children in the world – we just get a little older, the world gets a little larger, and the problems become all the more complex.

I wish you the best of luck in writing your story.

Respectfully yours,
Matthew A. Gallagher

APPENDIX

Here are various quotes, poems, and teachings for you to keep in mind as you continue on creating your path in life;

Life only makes sense looking backwards, but it must be lived moving forward.

A journey of a thousand miles begins with a single step.

And a journey worth living is not without its risks, pitfalls, and failures. Be willing and ready to learn with each step.

"To avoid criticism, say nothing, do nothing, be nothing." – Aristotle

"If you don't stand for something, you will fall for anything." – Malcom X

"It is not the critic who counts, not the man who points out how the strong man stumbles, or where the doer of deeds could have done them better. The credit belongs to the man who is actually in the arena, whose face is marred by dust and sweat and blood; who strives valiantly; who errs who comes short again and again, because there is no effort without error and shortcoming; but who does actually strive to do the deeds; who knows great enthusiasms, the great devotions; who spends himself in a worthy cause; who at the best knows in the end the triumph of high achievement, and what at the worst, if he fails, at least fails while daring, greatly, so that his place shall never be with those cold and timid souls who neither know victory nor defeat." –Theodore Roosevelt

Invictus – William Ernest Henley

Out of the night that covers me,
Black as the pit from pole to pole,
I thank whatever Gods may be
For my unconquerable soul.

In the fell clutch of circumstance
I have not winced nor cried aloud,
Under the bludgeoning of chance
My head is bloody, but unbowed

Beyond this place of wrath and tears
Looms but the horror of the shade
And yet the menace of the years
Finds and shall find me unafraid

It matters not how straight the gate
How charged with punishment the scroll

I am the Master of my fate
The Captain of my soul.

"Nothing in the world can take the place of persistence. Talent will not; there is nothing more common than unsuccessful men with talent. Genius will not; unrewarded genius is almost a proverb. Education will not, for the world is full of educated derelicts. Persistence and determination alone are omnipotent. The slogan 'Press on' has solved and will always solve the problems of the human race." – Calvin Coolidge

The Man in the Glass – Dale Wimbrow 1934

When you get what you want in your struggle for self.
And the world makes you king for a day,
Just go to the mirror and look at yourself
And see what that man has to say.

For it isn't your father, your mother, or wife
Whose judgment upon you must pass.
The fellow whose verdict counts most in your life…
Is the one staring back from the glass.

He's the fellow to please, never mind all the rest.
For he's with you clear to the end,
And you've passed the most dangerous, difficult test – if the man in the glass is your friend.

You may be like Jack Horner and "chisel" a plum,
And think you're a wonderful guy, but the man in the glass says you're only a bum,
If you can't look him straight in the eye.

You may fool the whole world down the pathway of years.
And get pats on the back as you pass,
But your final reward will be heartache and tears
If you've cheated the man in the glass.

The final, and most important thing to keep in mind is to remain true to your moral compass and values, because doing the right thing may not always be the easiest. In career, life, & love… there will be good days, bad days, and days. It's up to you to persist through the bad, maintain through the mundane, and always, always – champion the good.

"Darkness cannot drive out darkness, only light can do that. Hate cannot drive out hate; only love can do that." – Martin Luther King, Jr.

www.ingramcontent.com/pod-product-compliance
Lightning Source LLC
Chambersburg PA
CBHW020623300426
44113CB00007B/762